Saving Our World

ENERGY
CONSERVATION

Amanda Bishop

Marshall Cavendish
Benchmark
New York

Marshall Cavendish Benchmark
99 White Plains Road
Tarrytown, NY 10591
www.marshallcavendish.us

All Internet addresses were available and accurate when this book was sent to press.

Library of Congress Cataloging-in-Publication Data

Bishop, Amanda.

    Energy conservation / by Amanda Bishop.

    p. cm. -- (Saving our world)

    Includes bibliographical references and index.

    ISBN 978-0-7614-3224-1

1. Energy conservation--Juvenile literature.

2. Power resources--Juvenile literature.

3. Power (Mechanics)--Juvenile literature. I. Title.

    TJ163.23.B57 2009

    333.791'6--dc22

    2008006473

The photographs in this book are used by permission and through the courtesy of:

Half Title: Ana de Sousa/ Shutterstock; Oleg Prikhodko/Istockphoto.
Daniel Gustavsson/ Shutterstock: 4-5, Tim elliott/ Shutterstock: 6-7, ASSOCIATED PRESS: 8-9,
Chic Type/ Istockphoto: 10-11, Jose Gil/ Shutterstock: 12-13, Tony Latham Photography/ Getty
Images: 14-15, Francis Goussard/ Fotolia: 16-17, Getty Images: 19, Tomas Loutocky/ Shutterstock:
20, ilian studio / Alamy: 21, Ana de Sousa/ Shutterstock: 22-23, Ingvald Kaldhussater/
Shutterstock: 24-25, Alex Segre / Alamy: 26-27, Varina and Jay Patel/ Shutterstock: 28-29.

Cover photo: Daniel Gustavsson/ Shutterstock; ExaMedia Photography/ Shutterstock; Oleg
Prikhodko/Istockphoto.

Illustrations: Q2AMedia Art Bank

Created by: Q2A Media

Creative Director: Simmi Sikka

Series Editor: Maura Christopher

Series Art Director: Sudakshina Basu

Series Designers: Dibakar Acharjee, Joita Das, Mansi Mittal, Rati Mathur and Shruti Bahl

Series Illustrator: Abhideep Jha and Ajay Sharma

Photo research by Anju Pathak

Series Project Managers: Ravneet Kaur and Shekhar Kapur

Printed in Malaysia

1 3 5 6 4 2

CONTENTS

What Is Energy?

Energy is the power to do work. Every time you raise your hand, listen to a song on the radio, or ride on a bus, you use energy. Energy lights up your house at night, heats your water for washing, and keeps the food in your refrigerator cold.

What Is Conservation?

Conservation means using **resources** wisely. Conserving **energy** means using only as much energy as needed, and avoiding waste. When you jog around a track, you move at a steady, comfortable pace to conserve your energy so you can finish all the laps. Conserving is also important for other types of energy, such as **electrical energy**. We use electrical energy for tasks such as cooking, driving, and lighting our homes and schools.

Energy Conservation for Earth

Learning to conserve energy is important for our planet. Many of the ways we use energy today add **greenhouse gases** to the **atmosphere**. Greenhouse gases trap the Sun's heat near the Earth's surface. Trapped heat causes temperatures to rise. This phenomenon is called **global warming**. By using energy more wisely, we can help reduce greenhouse gas **emissions**.

How is energy used in North America?

Twenty-one percent of energy used in North America is used in private homes. Twenty-eight percent is used for transportation. Thirty-three percent is used to produce products. Eighteen percent is used by offices, schools, hospitals, and retail businesses. By making changes at home and on the road, and by learning about how your habits affect the rest of the world, you can contribute to energy conservation.

http://www.eia.doe.gov/kids/energyfacts/uses/consumption.html

Windmills use the energy of motion to create electrical energy. People use electrical energy in their homes.

Sources of Energy

Energy is all around us. Energy cannot be created or destroyed. However, energy can change forms. People change energy to make it work for them. The power that we rely on comes from energy that has changed forms.

Renewable Energy

All energy comes from **renewable** or **nonrenewable** sources. Renewable sources of energy keep providing energy no matter how much people use them. Renewable sources include **solar power**, or power from the Sun, and wind power. Nonrenewable sources of energy cannot be restored. Once people use them up, they are no longer available. Most of our nonrenewable energy comes from **fossil fuels**.

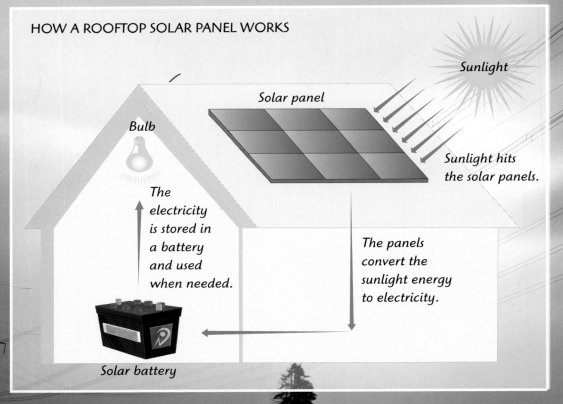

HOW A ROOFTOP SOLAR PANEL WORKS

Sunlight

Solar panel

Bulb

Sunlight hits the solar panels.

The electricity is stored in a battery and used when needed.

The panels convert the sunlight energy to electricity.

Solar battery

Power lines carry electrical energy from power plants to people.

EYE-OPENER

Coal, oil, and natural gas are called fossil fuels. They are the remains of ancient plants and animals that are buried deep in the earth. All living things are made in part of **carbon**. Over millions of years, the carbon in buried plants and animals has changed into fossil fuels. When fossil fuels are burned, they release their carbon as a gas called **carbon dioxide**. Carbon dioxide is a major geenhouse gas.

Depending on Fossil Fuels

For most of human history, people had to depend on renewable energy sources, such as windmills and watermills. Over the last two hundred years, however, we have come to depend more and more on fossil fuels for energy. Today, coal is burned in power plants to generate electricity. Oil is burned in the engines of cars, trucks, ships, and airplanes. Natural gas is used in furnaces to heat homes. More than 80 percent of the energy used in the United States today comes from fossil fuels.

Challenges and Changes

For a long time, the technology of power plants, vehicles, and homes was based on the use of fossil fuels. People now understand that fossil fuels can damage our environment. Renewable energy sources and the technology for using them are becoming more common. These sources are often more easily accessed than fossil fuels. They are also cleaner and less harmful to the environment. It can be expensive to switch to renewable energy at first, but a lot of money is saved in the long-term. Today about 6 percent of the energy used in the United States comes from renewable sources.

QUESTION TIME ?

What is energy efficiency?

Energy efficiency means wasting as little energy as possible. As energy changes forms, not all of it is used to generate power. For example, fuel in a car moves the parts, but it also generates heat that is of no use to the moving parts. Machines are energy-efficient if very little of the energy that is converted for use gets wasted.

THE COUNTRIES WITH THE LARGEST OIL RESERVES

Canada

Russia

Iraq

Libya

Iran

Venezuela

Nigeria

Saudi Arabia

Generating power can damage the environment in more ways than one.

Off the Road

North Americans rely on nonrenewable energy for transportation. Almost all vehicles are powered by fossil fuels. We can conserve energy by learning to think differently about vehicles—when, why, and how we use them.

Get Off the Road!

The best way to conserve energy in cars is to avoid using them. Whenever possible, ask whether your family might walk, ride their bicycles, or catch a bus or a subway instead of taking the car. If you are going a short distance and it is safe to go on your own, take your skateboard or scooter. You may have noticed that many major retail stores and malls are built outside of towns, which makes them difficult to reach without a car. Try to find local businesses that are more easily reached on foot or by bicycle. When groups of people need to go somewhere, such as school, activities, or work, carpool to reduce the number of cars on the road.

Running Cars Efficiently

Fuel efficiency is another important aspect of conserving energy. A fuel-efficient car uses as little fuel as possible to operate. New cars are usually more fuel-efficient than older cars. Small cars are usually more fuel-efficient than big cars or sport utility vehicles. If your family has more than one vehicle, conserve energy by using the smallest vehicle whenever possible. Light cars are usually more fuel-efficient than heavy cars, so make sure that there is no extra weight in the trunk or the bed of a pick-up truck. Properly inflated tires help a car run efficiently. Staying within the speed limit also helps reduce gas waste.

Crowded highways lead to greenhouse gas emissions and air pollution.

On the Lot

Fuel-efficient cars are now more affordable and available than in the past. Hybrid cars are among the most fuel-efficient cars on the road today. The technology for fuel efficiency is improving all the time.

Alternative Fuels

Scientists and engineers are working on new technologies for cars that will reduce our use of fossil fuels. **Hydrogen fuel cells** use hydrogen and oxygen to generate power without creating any pollution. **Biofuels**, such as **biodiesel** and **ethanol**, are made from natural substances, such as plant material or food waste. Ethanol, made from corn, was one of the first fuels used in cars in the early 1900s, but it was soon replaced by fossil fuels. Today, biofuel production and use is on the rise. Around the world, a few cities such as Seattle, and Stockholm, Sweden, have already introduced public transit vehicles that run on biofuel.

Hybrid cars use a combination of gasoline and electrical energy from a rechargeable battery.

Reduce emissions with carbon-offset credits. Many people travel by airplane for business or for vacations. Online groups help people determine the emissions caused by their travel. People can then buy "carbon offset credits" to make up for that amount of fossil fuel. This money is donated to help pay for projects that are good for the environment, such as renewable energy programs that reduce the use of fossil fuels.

The Cost of Transportation

To conserve energy, it is also important to think about your indirect fossil fuel use. Indirect use means that you do not use the fuel yourself, but you benefit from its use by others. For example, fossil fuel is used in trucks, airplanes, and ships to transport goods across states, countries, continents, and oceans. You benefit from the fuel these companies use to get you their products. Try to buy locally made or grown items instead. When you purchase goods, such as produce from local markets, you cut down on the energy used to transport the food from its source to your table.

Lights Out!

One of the main types of energy we use is electrical energy, or electricity.

Wasting Energy

Electricity is so readily available that we often forget to think about where it comes from. Most electricity comes from coal-fired power plants, which pollute and add greenhouse gases to the air. Coal, a fossil fuel, is burned to heat water. The water turns to steam, which moves **turbines** to generate electricity. Think about ways to use less electricity in your home. Turn out lights when you leave a room. Turn off all electrical devices that are not in use. Talk with your family about ways to use electricity more efficiently.

QUESTION TIME ?

How are compact fluorescent bulbs different from regular light bulbs?

Compact fluorescent light bulbs are more efficient than commonly used incandescent bulbs. Compact fluorescents last ten times longer and use about 75 percent less energy than do incandescent bulbs.

Drawing Power

Many appliances in our homes, including stereos, computers, and video game consoles are designed to be on **standby**. Even when these devices are switched off, they draw a small amount of electricity so that they will be ready to switch on as soon as you press the power button. Conserve energy by unplugging devices after you switch them off. To make it easier to remember, plug devices into surge-protected power bars. When you are finished using the devices, turn off the power bar and unplug it.

You can conserve electrical energy starting in your own home. Where else can you help conserve energy?

Indoor Energy

Many people in North America take their indoor environments for granted. Have you ever walked into a mall in the summertime and wished that you had brought a sweater?

Efficiency in the Home

Most furnaces run on natural gas, oil, or electricity. There are high- and mid-efficiency furnaces on the market today, but many older furnaces use a lot of energy to run. If your family needs to replace the furnace, ask your parents to find out about new technologies for efficiency. For example, some systems are designed to work with a hot-water tank to be as efficient as possible. Air conditioning units use electricity—a lot of it! By overusing air conditioning units, we contribute to global warming.

Heating and cooling the air inside is one of the biggest uses of energy in the United States.

Let the Sun Shine In

People sometimes forget that nature can help heat and cool homes. In cold weather, leave curtains open during the day to let sunlight warm your rooms. Close the curtains at night to keep the heat in. In warm weather, close curtains and blinds to keep rooms shaded from the sun. On warm days, open more than one window to allow a cross-breeze to carry cool air in and warm air out.

EYE-OPENER

Plant trees for temperature control. Often the west side of a house gets the most heat because of the direct sunlight during the late afternoons and evenings. Find out which side of your house, apartment, or school gets the most sunlight. Plant some trees or shade plants, such as climbing hydrangea, along that wall. The plants will help shade the house in summer and **insulate** it, or keep it warm, in winter.

17

Under Control

People can conserve energy by controlling the temperature inside their homes. The first step is to make sure that air stays inside your home. Avoid leaving windows open when the furnace or air conditioner is running. Many homes lose warm or cool air through windows and doors even when they are closed. Replace old windows with new windows designed for energy efficiency. Maintain windows by adding weather-stripping and caulking around the frames.

Programmable thermostats are not very expensive, and they can help save a lot of money in energy costs!

Get with the Program

Programmable thermostats let you set different temperatures for different times of the day. When everyone is out at work or at school, the temperature is automatically reset to conserve energy. A few minutes before everyone gets home, the programmed thermostat adjusts to make the temperature more comfortable. A programmable thermostat is especially important in the summer. Houses are at their hottest in the middle of the day, when the sun is shining. If no one is home, program the air conditioner so that it doesn't turn on till the end of the day, when people return home. It will save your air conditioner from working extra hard during the sunniest hours.

QUESTION TIME ?

What is the best temperature for a home?

In cold weather, the **optimum**, or best, temperature for your home is 68 degrees Fahrenheit (20 degrees Celsius). In warm weather, set the thermostat at 77 degrees Fahrenheit (25 degrees Celsius).

Proper insulation, weather-stripping, and caulking can help keep warm and cool air inside your home. Get expert advice on making the structure of your home more energy-efficient.

Hot Water

Heating water is a big drain on energy use. Heat moves toward the cold objects that surround it. Heat escapes quickly, so water has to be heated continually to maintain its temperature.

Grab a Quick Shower

By heating water efficiently and using hot water conservatively, families can reduce their energy consumption. Short showers require less energy and use less water than hot baths. Also, hot water in a bathtub loses its heat to the air very quickly, which may cause bathers to add more hot water to maintain the warm temperature.

Low-flow showerheads reduce the amount of water used in the shower.

Washing Up

People also use hot water for cooking and for washing dishes. When boiling water for warm beverages, only boil as much as you need. Keep lids on pots that are cooking on top of the stove. When washing dishes in the sink, pay attention to how much water you use, and see if you can be more efficient. Run the dishwasher only when it is full. Some new dishwashers are designed with two compartments that can be run separately to cut down on water use.

EYE-OPENER

Insulate your hot water tank to conserve energy. The insulation keeps heat from moving into the cold air around the tank, which is usually in the basement. If the hot water tank does not lose heat, it requires less energy to maintain the water's temperature.

Clean the Smart Way

One of our biggest uses of energy in the home is laundry. You have seen how much energy is required to heat water. Clothes dryers also use large amounts of energy. Talk to your family about conserving energy as you do laundry.

Washing Up

Can your family cut back on the number of loads of laundry done every week? Work out a laundry schedule, and make everyone responsible for getting dirty clothes to the laundry room on time. Washing clothes in cold water is an easy way to conserve energy because you don't use water from the hot water heater. Make sure that family members wash only full loads, and only clothes that need to be washed. When it is time to buy a new washing machine, remember that front-loading washing machines use less water and run more efficiently than top-loading machines. Look for the most energy-efficient washer and dryer you can find.

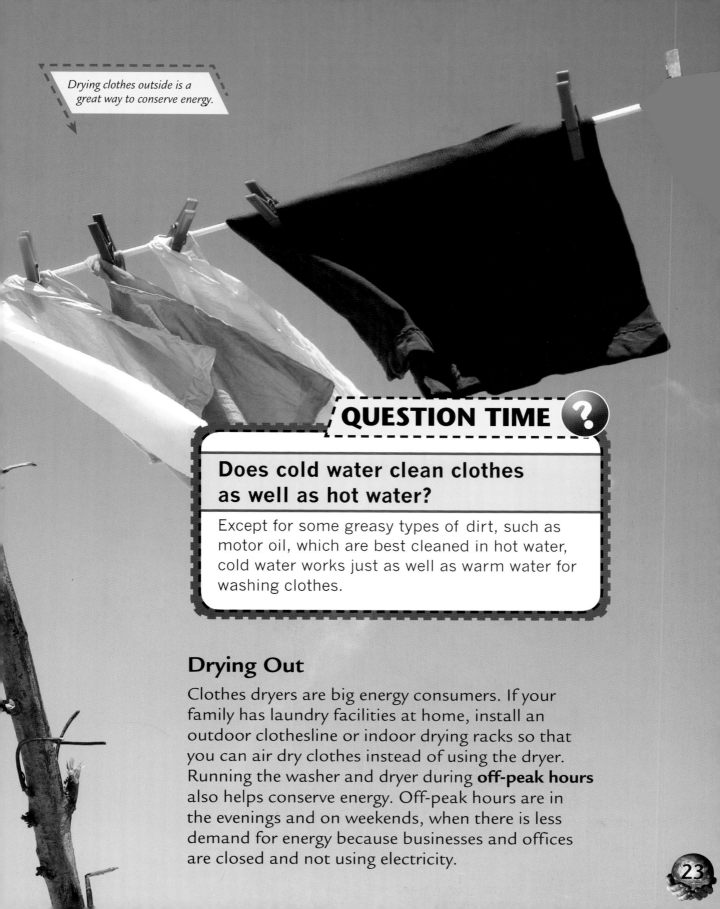

Drying clothes outside is a great way to conserve energy.

QUESTION TIME ?

Does cold water clean clothes as well as hot water?

Except for some greasy types of dirt, such as motor oil, which are best cleaned in hot water, cold water works just as well as warm water for washing clothes.

Drying Out

Clothes dryers are big energy consumers. If your family has laundry facilities at home, install an outdoor clothesline or indoor drying racks so that you can air dry clothes instead of using the dryer. Running the washer and dryer during **off-peak hours** also helps conserve energy. Off-peak hours are in the evenings and on weekends, when there is less demand for energy because businesses and offices are closed and not using electricity.

In the Kitchen

People use a lot of energy in the kitchen. Refrigerators, ovens, and dishwashers are big drains on energy. Small appliances, such as microwaves, toasters, and electric can openers also require a lot of energy to run.

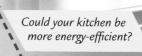

Could your kitchen be more energy-efficient?

Thinking about the Kitchen

Try to use the smallest available appliance for a task in the kitchen. Avoid using the oven for a job that can be done just as well in a toaster or microwave oven. Remember to unplug appliances that are not in use. Save energy by opening your dishwasher before it starts its drying cycle in order to let dishes air dry. Use the oven as soon as it is preheated. Do not let the fan over the stove run for any longer than necessary.

Keep It Cool

One of the biggest energy users in the kitchen is the refrigerator. Refrigerators work by maintaining a cool temperature. When food is put inside, the refrigerator draws out the heat in order to keep the internal temperature steady. Avoid standing in front of an open refrigerator deciding what to eat or pouring out a glass of juice. Let leftovers cool down for an hour before putting them into the refrigerator. The cooler food is when you put it in, the easier it is for the refrigerator to run efficiently. If there are coils at the back of your refrigerator, keep them free of dust so the refrigerator will run efficiently.

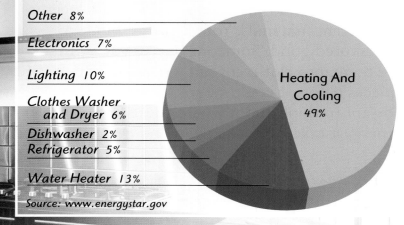

ENERGY USE IN THE HOME

Other 8%

Electronics 7%

Lighting 10%

Clothes Washer and Dryer 6%

Dishwasher 2%

Refrigerator 5%

Water Heater 13%

Heating And Cooling 49%

Source: www.energystar.gov

EYE-OPENER

Keep it simple. Many American families waste energy with little-used appliances in their basements, such as second refrigerators or chest freezers that are mostly empty. Freeze only as much food as you have room for in your kitchen freezer. Use cold cellars and garages to store cold drinks and preserves. If your family is ready to give up its second refrigerator, find out about refrigerator collection and recycling programs in your area.

Consume Less

Industries use energy to make goods and design technology. The goods and technologies are consumed, or used, by people like you. Becoming a smart consumer can help conserve energy.

Where Do They Come From?

People have all kinds of **manufactured goods** in their homes, from telephones to breakfast cereals. Manufactured goods are items that are put together from different materials. Your favorite shirt is a manufactured good. Someone had to gather the fibers, weave them into cloth, cut out a pattern, and assemble the pieces before it could be sold to you in a store. It is important to think about where goods come from and the energy it takes to make them and get them to you.

Reduce and Reuse

One of the best ways to conserve energy is to **consume** less. You can reduce your consumption by buying only what you need. Avoid products that are disposable or have a lot of packaging that will just be thrown away. Reuse items to reduce waste, since waste means more manufacturing. Reusable grocery bags and water bottles conserve energy used in industry. Shop for used clothing, furniture, DVDs, video games, and CDs at garage sales and secondhand stores. Donate your old goods to charity or hold a garage sale to keep your unused things in use.

EYE-OPENER

Recycle to conserve energy. Recyclable materials such as paper, aluminum, and plastic are collected from homes. They are then broken down for reuse in manufacturing. This process takes energy to complete, but the manufacturing of goods from recycled materials conserves energy. For example, paper made from recycled paper fibers takes as much as 70 percent less energy to produce than paper made from wood.

It takes a great deal of energy to ship goods from manufacturers overseas to chain stores in malls all over North America. Conserve energy by shopping in local stores for locally made goods.

Find Alternatives

Conserving energy is not difficult or inconvenient. It saves money on electricity bills. It can be somewhat expensive to buy energy-efficient cars, appliances, and light bulbs, but energy efficiency always adds up to savings in the long run.

Speak Up

Once you know how to conserve energy, it is easy to do. Tell your family, friends, teachers, and neighbors about easy ways to conserve energy. Keep investigating energy conservation in books, on the Internet, and in magazines. Every day, scientists come up with new ways for us to conserve energy and help save our planet. Learn about new developments. When you have a chance to make your voice heard or share with others, speak up!

Show by Example

Once you have made energy conservation a habit, you will find that there are all kinds of ways you can cut back on energy use. As others see what you are doing, they will start to think about energy use, too. Challenge your family or friends to activities that require no electricity or fossil fuel use. Ride your bike or skateboard, play catch in the park, or take pets for a nice, long walk. Turn off your television or computer for a few hours every night. Remember that nature has a lot of energy ready for you to use. Go for a swim at a public pool to cool off, instead of blasting the air conditioner. Bring your library book outside and read it under a tree instead of under a lamp!

Get outside and use your own energy to enjoy the outdoors!

Glossary

atmosphere: The body of gases that surround Earth.

biodiesel: Diesel fuel that is made up of organic material.

biofuel: Fuel that is made up of organic material.

carbon: A chemical element that is the basis of all living things.

carbon dioxide: A gas in the atmosphere that contributes to the greenhouse effect.

conservation: The act of using only what is needed.

consume: To take in, eat, or use.

electrical energy: Energy that is in the form of electricity.

emissions: The release of a gas into the atmosphere.

energy: The power to do work.

ethanol: A fuel made from the sugar in corn.

fossil fuels: Substances such as coal, oil, and natural gas that are made of the ancient remains of living things.

fuel efficiency: Using fuel in a way that wastes as little as possible.

global warming: The gradual increase of Earth's temperature due to greenhouse gases.

greenhouse gases: Chemicals in the air that trap the Sun's heat near the Earth's surface.

hybrid cars: Vehicles that use both gasoline and electrical energy to run.

hydrogen fuel cells: An alternative fuel source for vehicles.

insulate: To keep heat or cold contained.

manufactured goods: Items that are made by others for consumption.

nonrenewable: Describes a source of energy that does not replenish itself.

off-peak hours: Hours when energy consumption is lower because businesses and schools are closed, and therefore not using energy.

optimum: Describes the ideal or best condition.

renewable: Describes a source of energy that replenishes itself.

resources: A supply of something that can be used by people.

solar power: Electricity created using the Sun's energy.

standby: Describes appliances that draw power even when they are not turned on.

turbines: A machine that is made to rotate in order to generate usable energy.

Where to Find Out More

Visit these Web sites to find out more about energy and how you can help conserve it.

- This Energy Information Administration site,has all kinds great information on energy, as well as games and helpful glossaries.
  http://www.eia.doe.gov/kids/

- This U.S. Department of Energy site is a good one to visit with your parents. It has links to a home energy audit and a home energy check-up that can help you target areas in your home that could be made more efficient.
  http://www.energy.gov/forstudentsandkids.htm

- This site, hosted by the Environmental Protection Agency, explains what climate change is and how everyone can help fight it.
  http://epa.gov/climatechange/kids/difference.html

- This site, from the Office of Energy Efficiency and Renewable Energy in the U.S. Department of Energy, has tips, games, and helpful information.
  http://www.eere.energy.gov/kids/

- Visit this Alliance to Save Energy site to play interactive games and become a certified Energy Hog Buster!
  http://www.energyhog.org/childrens.htm

Index